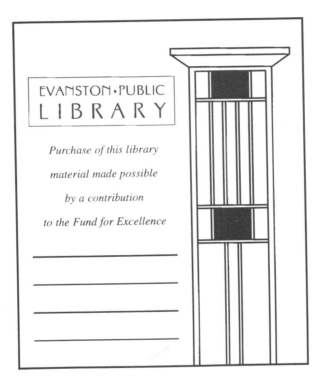

HISTORY

JOURNEY THROUGH TIME

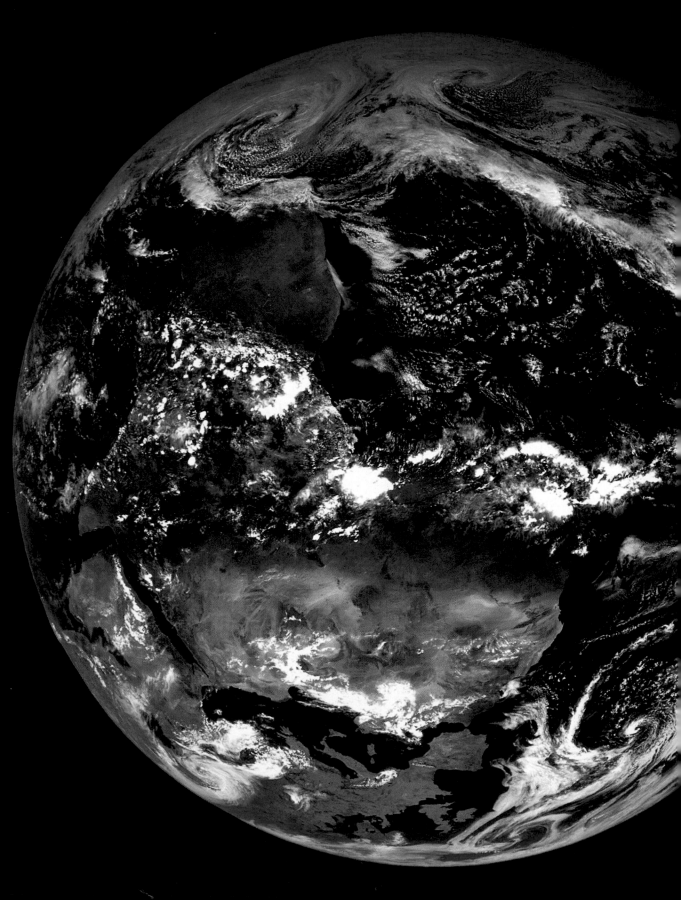

HISTORY

JOURNEY THROUGH TIME

ROY A. GALLANT

BENCHMARK BOOKS

MARSHALL CAVENDISH
NEW YORK

For Max

Series Consultants:

LIFE SCIENCES AND ECOLOGY

Dr. Edward J. Kormondy
Chancellor and (professor emeritus) of Biology
University of Hawaii—Hilo/West Oahu

PHYSICAL SCIENCES

Christopher J. Schuberth
Professor of Geology and Science Education
Armstrong Atlantic State University
Savannah, Georgia

Benchmark Books
Marshall Cavendish
99 White Plains Road
Tarrytown, NY 10591-9001

Library of Congress Cataloging-in-Publication Data
Gallant, Roy A.
History : journey through time / by Roy A. Gallant.
   p. cm. – (EarthWorks series)
Includes bibliographical references and index.
Summary: An overview of the history of Earth, the life that evolved on it,
and known periods of mass extinctions, from the planet's origin to the present.
ISBN 0-7614-1367-7
1. Historical geology—Juvenile literature. [1. Historical geology.] I. Title.
QE38.3 .G35 2002
551.7—dc21   2001043253

Photo research by Linda Sykes Picture Research, Hilton Head, SC

Cover: Carr Clifton/Minden Pictures
The photographs in this book are used by permission and through the courtesy of: Photo Library
International/Science Photo Library/Photo Researchers: 2–3; David A. Hardy/ Science Photo Library/ Photo
Researchers: 6; Material created with support to AURA/STScl from NASA/ Space Images: 10; Francois
Gohier/Photo Researchers: 20, 29, back cover; Scott Camazine/M.Marchaterre/Photo Researchers: 22; A. B.
Dowsett/Science Photo Library/Photo Researchers: 23, back cover; Jim Steinberg/Photo Researchers: 24–25;
Martin Land/Science Photo Library/Photo Researchers: 30; John Cancalosi/Peter Arnold: 31; Layne
Kenney/Corbis: 32; Vaughn Fleming/Science Photo Library/Photo Researchers: 33; Tom McHugh, Field
Museum, Chicago/Photo Researchers: 34–35, 36–37, 42–43, back cover; Ludek Pesek/Science Photo
Library/Photo Researchers: 44–45; American Museum of Natural History: 48, 51, 58 (top), 58–59, 60; Mark
Smith/Photo Researchers: 54; Photo Researchers: 56-57; J. Sibbick/Natural History Museum, London: 59
(top); Mark E. Gibson, 64–65; Elizabeth Weiland/Photo Researchers: 66; Jonathan Blair/Corbis: 66–67; E. R.
Degginger/Photo Researchers: 70–71, back cover.

Series design by Edward Miller.

Printed in Hong Kong

6 5 4 3   2   1

Title page: As photographed in space by the European Meteosat satellite, Africa's sprawling Sahara
desert is the dominant central feature (in brown). Above it is tiny Italy and the rest of Europe, which
looks green because of its vegetation.

CONTENTS

INTRODUCTION

To unravel Earth's long and tangled history we must go back in time, nearly five billion years, when the Sun and its family of planets were just forming out of a great cloud of cosmic gas and dust. We have to go back that far because the materials out of which Earth was made were being assembled in ways that would later determine whether life could arise and prosper on the new planet's surface.

It was a dramatic time in the history of our planet. Countless volcanoes were erupting, bubbling and sputtering the gases, ash, and dust that became a primitive atmosphere. More gases hissed and roared out of countless geysers and helped provide water for the first ponds and seas. Meanwhile, giant comets of loosely packed ice and dust were to add more water to the planet, making Earth the only place in the Solar System to have large bodies of the precious liquid.

Still later, energy from the Sun helped stir the chemical soups of shallow seas and ponds, sparking the formation of small clusters of atoms called *molecules* that became the complex chemical "seeds" of the first primitive living matter. These molecules also were carried to the planet by meteorites and comets.

VIOLENCE AND CHANGE

However life started on Earth, once it gained a hold it took off in a grand style, growing more abundant and more complex. Along the corridors of time it evolved into the magnificent and bewildering variety of plants and animals that have come and gone or persist to this day. Some biologists say

In its violent youth, Earth was a chaos of erupting volcanoes that poured rivers of lava over a landscape bombarded by comets, asteroids, and gigantic planetesimals. The battered planet was a molten mass for hundreds of thousands of years. Eventually Earth cooled and formed a solid crust that became bathed with ponds, rivers, and seas.

that life on Earth has been so abundant since the planet's crust became solid nearly four billion years ago that 99 percent of the plants and animals that have ever lived are now gone.

We know of five periods of mass extinctions that have wiped out anywhere from 50 to 90 percent of all living things. And there may have been as many as 20 such periods, some caused when gigantic sheets of ice up to 2 miles (3 kilometers) thick covered parts of the planet. Other mass extinctions were caused by comets and asteroids smashing into the planet and filling the air with so much dust and debris that sunlight was seriously blocked and the lives of many animals and plants snuffed out. Widespread volcanic eruptions also must have contributed to some extinctions.

It was over only the last 550 million years or so that most of Earth's life-forms arose, flourished for a while, then gave way to other organisms better able to take advantage of changes in the environment. As life evolved, so did the face of the planet. Inland seas dried up and turned into deserts. The land was continually altered, sculpted by ever-changing climates. By a process called plate tectonics, restless forces within the planet shifted the continents around like giant rafts floating on a sea of molten rock. Mountain ranges were thrust up, worn down, and formed anew while deserts were flooded to become oceans. Glaciers ground their way across the land, crushing everything in their path. As the giant blocks of ice melted from time to time, sea levels rose by hundreds of feet.

Although Earth seems stable and to persist with little or no visible change, flux has always been the rule. It has been the driving force that has continually shaped and reshaped the planet, and it continues today. But the geological clock ticks so slowly we hardly notice. What an experience it would be if we could somehow tinker with that clock, turn it back, and witness Earth's history unfold before our eyes.

ONE

A BEGINNING
FOR EARTH

The Solar System, our cosmic home, is an orderly community. It contains one star—the Sun—and an assortment of planets and their moons. Additionally, there are millions of chunks of rock and metal called asteroids, and millions more of those iceballs known as comets. All are held as gravitational captives of our local star. One of those objects is planet Earth, our geographic home.

We think that the stars and their planets formed out of huge clouds of gas and dust called *nebulae*. We can see nebulae in many parts of space. Their gas is mostly hydrogen along with some helium, and their dust is made up of tiny bits of solid matter.

The Sun and planets were born about 4.6 billion years ago out of one such nebula. The cloud stretched at least 19 billion miles (30 billion kilometers) and contained about twice as much matter as the Sun has today. Gravity, which was strongest in the dense central part of the cloud, pulled matter in from the outer regions. Eventually the cloud's gravity caused it to collapse in on itself. As it did, the matter in the center was packed tightly

Star birth galore goes on in those uncountable cosmic clouds of gas and dust called nebulae that are part of galaxies. The youngest stars in this cloud, known as NGC 4214, some 13 million light-years away, are still encased in their cocoons of hot gases. They are seen as five large white patches at the lower right. At center is a cluster of hundreds of massive blue stars 10,000 times brighter than the Sun.

together, all the while heating as the pressure there slowly increased. The continued inrush of matter also caused the cloud to start spinning and spreading out to form a large whirling disk called a protoplanetary disk. About 90 percent of the cloud's gas and dust became part of a sphere assembling at the disk's center. Over time, this globe of gases grew hot enough to glow with a dull red light.

Within the great wheel of disk material stretching out from the Sun's equator, tiny dust grains were drawn together and began forming clumps, small at first but then growing into larger masses, in a process called *accretion*. Some were ices, others were rocky material, and still others were heavier matter including iron and other metals. They grew to various sizes—depending on how much matter they swept up—becoming the planets and their major moons we know today. After about 100 million years, Earth had gathered perhaps 98 percent of its present amount of matter. The main ingredients of the disk matter out of which Earth formed were hydrogen, helium, carbon, nitrogen, oxygen, silicon, iron, nickel, aluminum, gold, uranium, sulfur, and phosphorus.

As the young Sun continued to heat, its glow changed from a cherry red to the hotter yellowish white glare we see today. As more and more of the nebula's hydrogen gas and dust tumbled into the core, the Sun became so hot that its hydrogen atoms fused and formed the heavier element helium. In the process, tremendous amounts of energy began to be released. We see that energy as light and feel it as heat.

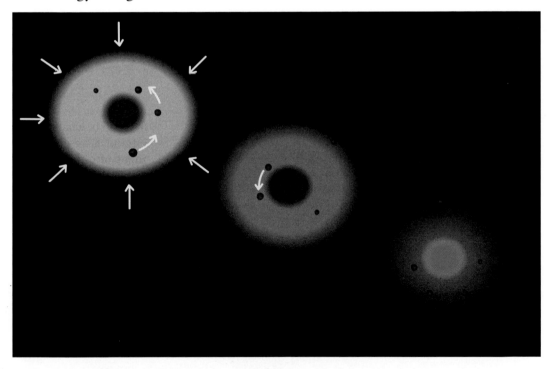

The Sun and planets were condensed out of a great cloud of gas and dust of a nebula. Gravity caused most of the nebula material to pack into the center of the cloud, heat up, and begin to glow as our young Sun. Other matter of the spinning disk of material formed "seed" planets that swept up more and more material that became the planets we know today.

EARTH GETS A SOLID CRUST

Earth was once a soupy ball of melted rock and metals that glowed red at a temperature of more than 3600 degrees Fahrenheit (2000 °C). Some of this warming was caused by clumps of matter, called *planetesimals*, striking the young planet. Still more heating came from certain heavy atoms breaking apart and releasing energy in the form of radioactivity. The cosmic icebergs that rained down onto the hot planet were instantly turned to steam that shrouded Earth in a cocoon of water vapor.

At one point, according to some astronomers, a massive planetesimal the size of Mars or larger collided with Earth and splashed matter into surrounding space. Within 10 hours, most of this rocky debris had formed into a sphere some 1,200 miles (1,930 kilometers) across, which then cooled and solidified as the Moon.

EARTH'S STRUCTURE

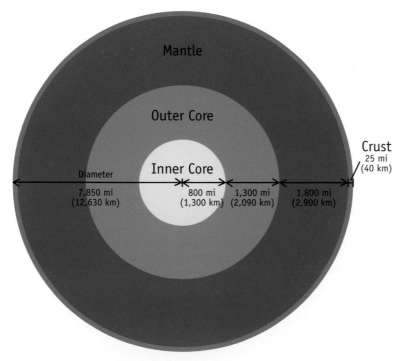

Geologists who study Earth's interior have discovered zones of different kinds of materials. At the surface is a thin layer of lightweight crustal rock. Deeper down is an enormous zone of much denser rock—the mantle. The planet's central region consists of an outer core of liquid iron-nickel within which is a smaller core of solid iron-nickel. (Radius distances shown in this diagram are approximate only.)

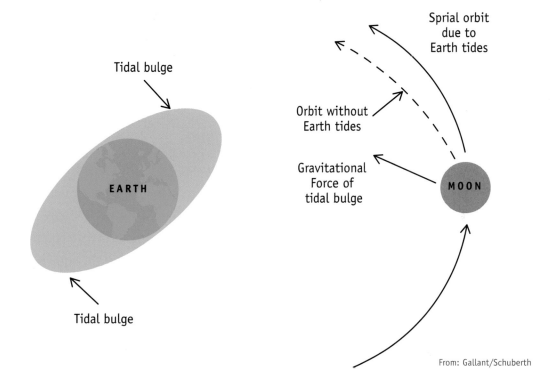

Tidal bulge

EARTH

Tidal bulge

Sprial orbit
due to
Earth tides

Orbit without
Earth tides

Gravitational
Force of
tidal bulge

MOON

From: Gallant/Schuberth

Earth's rotation drags its tidal bulges raised by the Moon ahead of the direct path from the Moon to Earth. The mass of those bulges pulls the Moon forward and ever so gradually whips it along an outward spiral.

Eventually, the heavy rain of planetesimals slowed as more and more of them were swept up by the young planet. This allowed Earth's molten rock at the surface to cool and become solid in places. Still, from time to time huge planetesimals smashed into the solid crust, breaking it in places into free-floating masses of rock suspended on the sea of molten rock below.

Young Earth's molten rock and metal material gradually settled and separated. While the heavier matter, such as iron and nickel, sank into the core, the lighter silicate rock materials floated to the surface. As the crust cooled some four billion years ago, there were still many asteroid-sized planetesimals crashing into Earth and the Moon. The thousands of lunar craters are evidence of this bombardment. Many asteroid strikes left enormous craters in Earth's crustal rock, some of which are still visible today. As they did, huge amounts of molten rock welled up and spilled out over large areas. Because the crust was no longer molten when these late-arriving planetesimals struck, the heavy metals in them remained as part of Earth's crust instead of sinking into the iron and nickel core.

No one can be sure what the landmasses looked like after Earth's crustal rock solidified. But they hardly resembled the major continents we know today. We can trace the evolving shape of the land back to a little more than 250 million years. Around that time, there was a supercontinent called Pangaea. By about 150 million years ago, Pangaea had broken apart into a large northern continent called Laurasia and another sizable southern continent called Gondwana. By about 65 million years ago, those giant landmasses subdivided even more, taking on the shapes we recognize as continents today.

In the early 1900s, the German scientist Alfred Wegener imagined that some 250 million years ago an enormous supercontinent called Pangaea made up Earth's surface. Over time, Pangaea was broken apart by crustal movements into two lesser continents, one in the north called Laurasia and one in the south called Gondwana. By about 65 million years ago, the continents had broken apart even more into just about the positions they are in today.

Our Lunar Legacy

When Earth acquired its one natural satellite, the Moon, something odd happened and continues to happen to this day. Time slowed down. Or to put it another way, Earth's days began to grow longer. So far as we can tell, when the Moon was torn out of Earth's side, Earth was rotating, or spinning, on its axis so fast that the length of a day was only five hours instead of twenty-four. But with the formation of seas and the rising of ocean tides caused by the Moon's gravitational tug, things started to change. Water dragging against the ocean floors acted as a brake that slowed Earth's *rotation*. Measurements over the past three hundred years show that the slowing of Earth's rotation was making the days longer by 0.002 seconds a century. Although that is not very much over a few years or centuries, it added up over the next 4 billion years or so. By about 570 million years ago, a day was already 21 hours long.

The ocean tides are also causing something even more interesting to happen to our Earth-Moon system. The tidal bulge of water raised by the Moon is nudging the Moon to spiral slowly away from Earth. The gravitational tug of this mass of tidal water speeds the Moon forward in its orbit and so causes the orbit to spin out in an ever-widening path. Measurements by laser beams show that the Moon is retreating from Earth at a rate of about 1 inch (3 centimeters) a year. That will increase the Moon's distance from Earth by some 16,000 miles (25,750 kilometers) over the next million years. Eventually, the slowing of Earth's rotation will cause the then distant, and therefore smaller-appearing, Moon to hang motionless in the sky, no longer rising and setting each lunar day. When that time comes, some people will witness the Moon in the same place in the sky every night while people living on the opposite side of Earth will never see it at all.

An Atmosphere and Seas

During Earth's early stages, many gases were released and collected above the new planet as a primitive atmosphere. Among them were large amounts of hydrogen, helium, water vapor, nitrogen, carbon monoxide, carbon dioxide, and smaller amounts of methane, ammonia, and hydrogen sulfide. The air was also heavy with poisonous cyanide and formaldehyde, and there was little or no oxygen. As more and more water vapor collected in the atmosphere, the temperature of the air eventually dropped, allowing the vapor to condense and fall as rain. In some areas, where the surface rock was cool enough, the rain soaked into the dry rock. In other places, where the rock was still very hot, the rains boiled, evaporated into water vapor, and were sent back into the atmosphere.

But as the crustal rock continued to cool, torrential rains fell day and night for perhaps one hundred thousand years. They collected in pools that eventually formed shallow, warm seas. By about 3.9 billion years ago, a thin, solid crust wrapped almost all of the planet. From time to time this thin crust continued to be punctured by eruptions from below, and molten rock welled up, evaporating some seas and melting huge areas of solid crust. This part of Earth's history has been appropriately called the Hadean (meaning "hell-like") stage. It is hard to imagine even the slightest glimmering of life beginning on our planet then, but after the Hadean, life was about to make its debut.

TWO

A BEGINNING FOR LIFE

The most important event in Earth's history, after its formation as a planet, was the appearance of life. But how did it happen? Did it begin here at home, or did the seeds of life, in the form of complex chemicals, drift to Earth from outer space and so give rise to the great chain of organisms that have helped shape its history?

ENERGY + MATTER = LIFE

To begin our inquiry into the origins of life, we must turn our clocks back 3.9 billion years. At that time Earth's surface was changing, and so was its atmosphere. Energy from the Sun was turning ammonia (NH_3) gas into free hydrogen and nitrogen. Methane (CH_4) was broken down into carbon and hydrogen. Water vapor (H_2O) was being separated into its essential parts—free hydrogen and oxygen. The scene was then set for what biologists now think were the early stages of the origin of life.

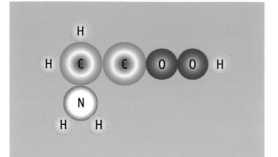

Our bodies contain about twenty different types of amino acids, the essential chemical building blocks of life. The simplest one is called glycine. It is made up of two atoms of carbon, two of oxygen, five of hydrogen, and one of nitrogen.

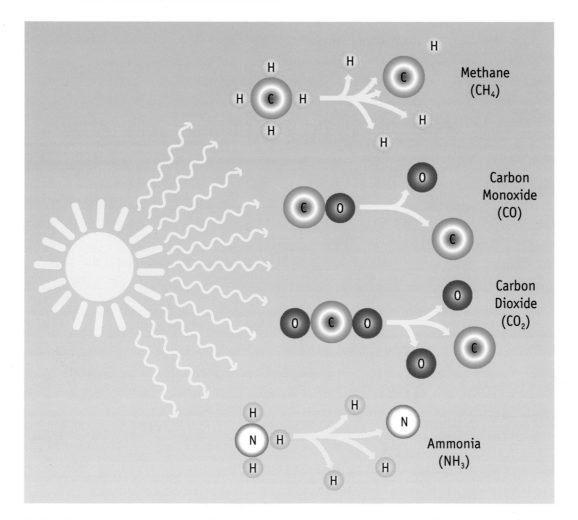

Methane (CH₄)

Carbon Monoxide (CO)

Carbon Dioxide (CO₂)

Ammonia (NH₃)

In Earth's early atmosphere, the Sun's energy broke down methane into free carbon and hydrogen atoms, and carbon monoxide and carbon dioxide into free oxygen and carbon atoms. Ammonia was broken down into free hydrogen and nitrogen atoms.

Many biologists think that life could have arisen in Earth's early warm seas. Energy from the Sun and from lightning, for example, can cause such common elements as carbon, oxygen, nitrogen, and hydrogen to join and form the complex clusters of atoms, called molecules, needed for life. Among these molecules are certain chemical building blocks called *amino acids*. Amino acids combine with each other to form still larger clusters called *proteins*. Our bodies use proteins for growth, for the repair of injured parts, as a source of energy, and to fight disease. Amino acids also come ready-made from space, since they have been discovered in meteorites.

But how could nonliving molecules, no matter how complex, become living matter? Here the problem may be in the words we use, as we usually describe things as either living or nonliving, but with nothing in between. Instead we should be thinking of an unbroken chain of chemical forms ranging from elements to simple molecules to complex clumps of molecules. Eventually systems of molecules would develop that had certain things in common with the matter we call "living."

Once amino acids and proteins were appearing in the warm seas, the next stage in the rise of living matter could begin. This arrived in the form of complex groupings of molecules that were enclosed within protective jackets called cell membranes. The membrane of each cluster acted as a sac that separated the outside environment from the environment within. However, tiny holes in the membrane let certain food molecules enter the sac. These were used as building blocks for growth and as a source of energy. The holes also let waste matter made by the proteinlike cluster pass into the outside environment. We can look on those membrane-bound clumps of complex molecules as a primitive form of living matter.

We can then imagine certain of these clusters acquiring a new ability that made them more advanced and better equipped for survival. Instead of taking in existing food molecules, they collected other, smaller molecules that they then assembled into food within their membrane sacs. Since there would be more of the smaller molecules in the outside environment than

the ready-made portions of food, those proteinlike clumps able to make their own meals would have an advantage over those that could not.

Biologists imagine some such chemical processes eventually producing the first living *cells*, which are the smallest organized units of living matter. Among the oldest cells we know of are some from Greenland that may have lived as long as 3.86 billion years ago. Cells were the organisms that became expert at using the raw materials in the outside environment to sustain themselves. Those raw materials were carbon dioxide and water vapor of the air. With sunlight as a source of energy, the cells combined these building blocks and made a sugar called *glucose*. In the process, called *photosynthesis*, they released oxygen, as "waste" matter, into the air.

There are many gaps in our knowledge of the chain of chemical events that led to the creation of the simplest forms of life. Even so, most biologists agree that a process such as this led to the formation of the first cells. They also generally agree that with the right conditions—that is, the right blend of energy and matter—life of some sort is bound to arise. This is true not only for Earth but for countless other planets throughout the Universe as well.

THE AGE OF BACTERIA

Bacteria are not only the simplest cells we know of but most likely also the oldest. There are trillions of them, everywhere. They drift through the air, live in the ooze of lake bottoms, thrive on the snow of Antarctica, and dwell in rock deep beneath the ground. They hitch rides on particles blown across the ocean from Africa and China during raging dust storms. Some bacteria are deadly germs, but many others are helpful. Billions of bacterial cells make your body their home where they help digest your food, for example.

Looking like small, rounded boulders, stromatolites are half-dead, half-living communities of countless bacteria. They are the survivors of Earth's first living organisms, which thrived more than 3.5 billion years ago. Among the few that remain are these along the shore of Shark Bay in Western Australia.

Their ability to reproduce is astonishing. Where there was only one before, twenty minutes later there are two. And twenty minutes after that there are four. In only a day a single bacterium can give rise to a million new cells just like itself.

The biologist Lynn Margulis tells us that there was a time in Earth's history during the Proterozoic era some 2.5 billion years ago, which she calls the Age of Bacteria. Broad and colorful blankets of bacteria stretched from horizon to horizon and painted the land, lakes, rivers, and shallow seas shades of green, brown, yellow, and purple. For the first few billion years, she says, bacteria may have been the only living organisms on Earth. Eventually, though, they gave rise to all other life forms.

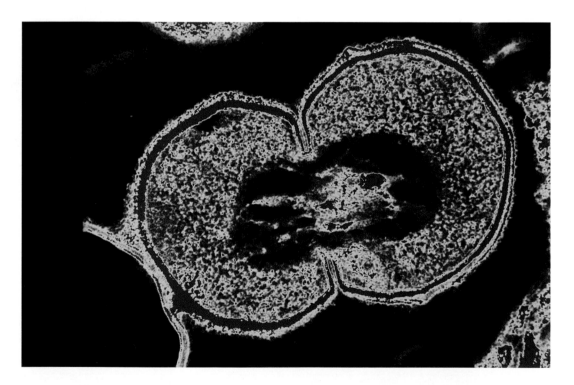

Bacteria make more bacteria by dividing in two. Above, a bacterial cell is beginning to divide by pinching itself in half. At right, a bacterial cell called E. coli, which lives in your intestine, has just about completed its division. Its nuclear material is at the end of each new cell. It takes a single bacterium about twenty minutes to produce two new bacteria.

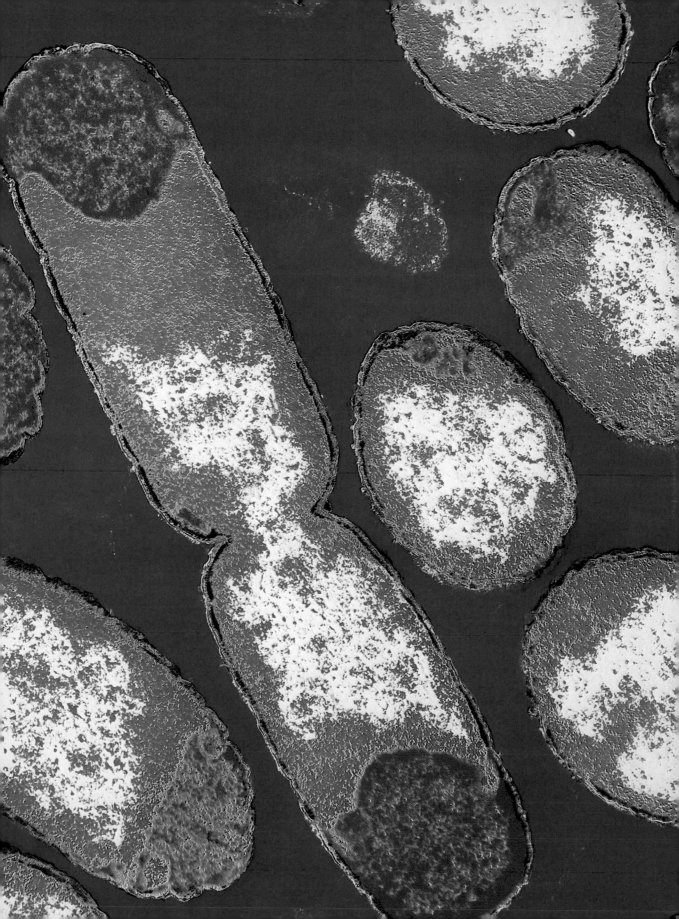

THE OXYGEN REVOLUTION

Some of the bacteria colorfully adorning the landscape some two billion years ago developed the ability to make their own food through photosynthesis out of hydrogen and carbon dioxide in the air in a process called *photosynthesis*. As they did, they began pouring huge amounts of oxygen into the air. What seemed like a milestone in the history of life brought on the greatest natural catastrophe that Earth's organisms had ever experienced.

At first the oxygen was quickly absorbed by iron and certain other oxygen-hungry elements in the seas as well as in other parts of the environment. After the nonliving parts of the environment had absorbed all the oxygen they could, the surplus remained in the air as free oxygen. That's when the trouble started.

Free oxygen quickly combines with and then breaks down the complex molecules of living matter. It breaks down vitamins. It destroys proteins and cell membranes, killing a cell instantly—unless the cell has a built-in protection against the poisoning gas. During the Age of Bacteria, few cells had evolved any such safeguard. As a result, many populations

Colorful bacteria decorate several of Yellowstone National Park's hot springs. The bacteria serve as thermometers because those of certain colors live within certain ranges of water temperature.

of ancient bacteria were wiped out during the oxygen revolution. Those that were not killed found ways to thrive in the new oxygen-rich environment. Among them were the blue-green algae called *cyanobacteria*, which survive to this day. They form the scum on swimming pools and coat our shower curtains. Other types of ancient bacteria coped by taking up life in the mud of lake and river bottoms where there wasn't any oxygen to harm them.

It was the survivors who adapted to the new environment that thrived and evolved into more complex cells. And it was those complex cells that over the hundreds of millions of years to come branched off into the staggering variety of higher life forms revealed in the fossil record.

The span of time we have been exploring so far in our brief account of Earth's history is called the Precambrian, which fills out 80 percent of Earth's history to date. The Precambrian was followed by the Phanerozoic eon, which began some 540 million years ago with the Cambrian period and continues to this day. It saw the explosion of a staggering number of new life forms.

THREE

THE PAGEANT OF LIFE BEGINS

To keep track of Earth's past, scientists refer to the geologic calendar. The fossil record that provides traces of these various ages continues to offer mountains of information about the life forms that have lived over that past half billion years or more and how they have changed through time in a process called *evolution*. Analysis of fossil plants and animals as well as Earth's rocks gives us insights into the ups and downs of the planet's climates over the ages. Although our understanding is incomplete, each year more information is discovered that adds new pieces to the grand puzzle of Earth's pageant of life.

The first cells had some three billion years to develop into more complex cells, then into the colonies of cells that formed new organisms of all kinds. Three billion years is a very long time. During that unimaginable span of years, centuries, and millennia, evolution had limitless opportunities to give birth to an equally unimaginable array of animals and plants.

According to the fossil record, soon after the Cambrian period began about 540 million years ago, there was a bewildering variety of ocean plants and animals in such abundance that biologists are quite puzzled. No one has yet convincingly explained how that explosion of new life-forms came about, although many have theories.

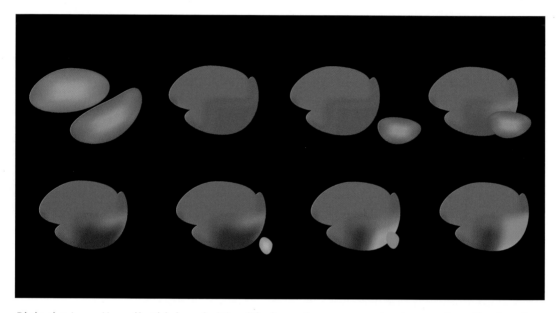

Biologist Lynn Margulis thinks primitive life forms became complex by merging. She imagines that two cells, each with a specific talent, became a single cell that inherited both skills. And the two-talent cell picked up a third, and then a fourth, and so on.

One theory, called "snowball Earth," was championed by Harvard geologist Paul Hoffman. The snowball Earth theory is that, for some 10 million years before the Cambrian period, Earth was gripped in a planetwide ice age. During this time, life died down to almost nothing. Then, over a few million years, volcanoes released large amounts of carbon dioxide into the air. As the gas built up, it began to act as a greenhouse and the planet warmed enough to melt the ice and form huge shallow seas. Life then took a new hold, abounded in the new environment, and gave rise to the Cambrian explosion of new life forms.

How Fossils Are Formed

Earth's rocky crust is a vast graveyard that contains the fossil remains of plants and animals that have lived throughout most of Earth's history. Fossils are the facts of life's existence traced back through geologic time to nearly four billion years ago.

The seafloor is one of many places where fossils are formed. When a fish or other marine animal dies, its remains drift to the bottom where eventually they may be covered by an undersea mudslide. Over millions of years, the mud changes to rock, which may later be thrust up as a mountain.

The wood of this petrified log found in Arizona's Petrified Forest National Park is from the Triassic period. It was completely replaced millions of years ago by mineral-bearing water that seeped down through the sediments and was soaked up by the long-buried tree. Gradually the wood dissolved and only the hardened minerals that replaced it remained.

Or the sea may dry up, exposing its rock floor. Erosion by wind and rain then wears away the rock and exposes the fossilized animal remains.

Soft-bodied animals such as jellyfish stand a poor chance of becoming fossils. Bone, shell, or hard wood, which don't decay very quickly, stand a much better chance of becoming fossils. Sometimes almost an entire animal has been fossilized by being frozen in ice more than 25,000 years ago.

Many of the plants and animals that are now part of the fossil record were preserved because their bone, shell, or other hard parts were changed into a different substance. Mineral-bearing water slowly seeping down through clay, mud, sand, or other sediments was soaked up by the porous bones, shell, or wood. As the water gradually evaporated, the

Above: A cut and polished ammonite fossil that is more than a hundred million years old was preserved by mineral replacement. Right: This fossilized plant, a horsetail, was common during Earth's Carboniferous period.

Many sea lilies (crinoids) flourished during the Ordovician period, They were attached to the seafloor by flexible stalks of calcite encased in flesh.

minerals left behind filled the small open spaces within. Brightly colored silica, calcite, or orange and red iron compounds often become part of a fossil. Frequently the bone or shell is dissolved by the groundwater and is slowly replaced by its minerals. In some petrified wood, silica has not only filled in small hollow spaces, it has replaced the once-living woody tissue. This has happened so perfectly that the individual cells and annual tree rings show up exactly as they once appeared, only millions of years later.

Fossils are our keys to past life and conditions on Earth. They clearly reveal in fine detail the many chapters of Earth's history.

The fly (left) and mosquito (right). In ages past, insects such as these were trapped in oozing tree resins that hardened into stonelike beads called amber.

THE CAMBRIAN PERIOD

During the early Cambrian period, shallow seas covered the edges of the landmasses. Later in the period the seas spread inland. All the while sediments were being washed into those seas, sometimes filling extremely long, shallow underwater depressions. Millions of years later, the materials in one such depression were to be thrust up as the Appalachian Mountains. More recently the Rocky Mountains were created in a similar manner.

Many new life-forms had emerged by the beginning of the Cambrian. Cambrian fossils have been found in the eastern United States, Canada, and Wales, for example. Land animals and plants had not yet evolved, nor had *vertebrates*, animals with backbones. The marine animals that had developed included sponges, trilobites, brachiopods, graptolites, and others without backbones, known as invertebrates. Trilobites were especially plentiful, with thousands of different types, or *species*. Some swam, others crawled, and still others burrowed into the muddy seafloor. All had hard outside skeletons with their backs divided into three segments by furrows. Most species were only 1 to 3 inches (3 to 7 centimeters) long. About 70 percent of all Cambrian fossils are trilobites.

A mid-Cambrian period seascape of about 520 million years ago.

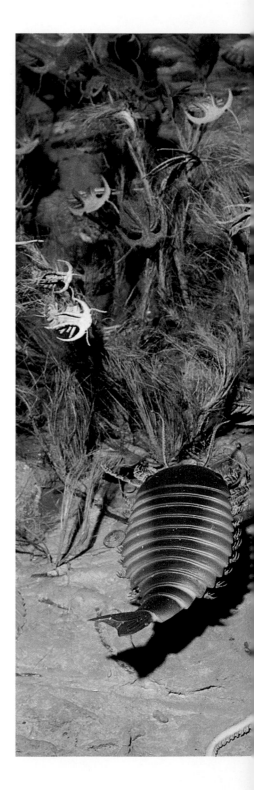

The early Cambrian climate seems to have featured long cold spells later followed by milder periods. By the end of this time, almost all of North America had been flooded and so abounded with Cambrian life.

THE ORDOVICIAN PERIOD

The next geologic period began about 500 million years ago. During this time most of the Northern Hemisphere was still flooded with shallow seas. So the Ordovician continued to be rich with marine life. Fishes without jaws appeared during the middle of the period. Trilobites reached their greatest numbers, as did graptolites, bryozoans and brachiopods. Graptolites were tiny animals that lived in colonies. Bryozoans are small animals that live attached to the seafloor; many thrive in today's seas and include mosslike organisms, others shaped like fans, and still others in the form of thick stems. Some brachiopods exist to this day as well, inhabiting mud flats around the world and look almost exactly like their Cambrian

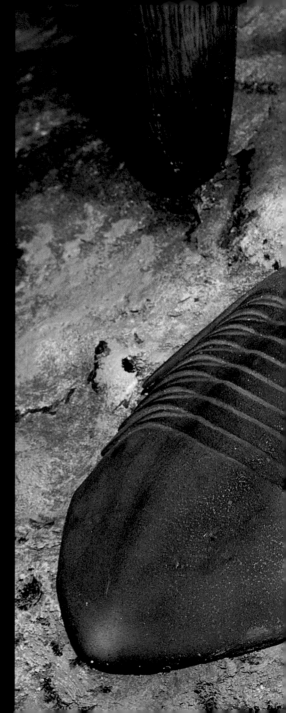

A late Ordovicaian period seascape of a trilobite dating about 450 million years ago.

ancestors. These animals have two shells, like those of a clam. They lived attached to the bottom of the seafloor and captured food that drifted by with small hairlike tentacles. The first corals emerged during this period as well.

Among the most important Ordovician fossils ever found are some discovered in Colorado sandstone. They date from about the middle of the period and are the earliest known remains of animals with a backbone. They may represent the first vertebrates, who were the ancestors of all of today's mammals, birds, reptiles, amphibians, and fishes.

THE SILURIAN PERIOD

This period began about 440 million years ago and was a time of great volcanic activity in what is today northeastern North America. Extensive layers of salt from dried-up seas were also deposited in western New York and Michigan, for example. The Atlantic Ocean didn't exist at this time, so you could have walked from Canada to Europe. The shallow Silurian seas abounded with giant sea scorpions, called eurypterids, some up to 10 feet (3 meters) long. The trilobites, so successful for 150 million years, began to die off. Most successful among Silurian animals was a group with hard outer skeletons, like today's lobsters, called *arthropods*. Some swam or burrowed, while others took to the air. All had bodies divided into segments.

Still other organisms had developed thick, protective scales. Near the end of the Silurian new types of backboned creatures were evolving. These were the placoderms, fish that were armor-plated and had jaws capable of biting, cutting, and crushing.

Perhaps the most important event during the Silurian was that for the first time life took hold on land, and stayed there. The first land plants may have been mosses, followed by psilopsids, relatives of ferns, which still exist. Late in the period animal life also moved out of the water. Millipedes and scorpionlike animals were among the earliest land dwellers. They most likely fed on the remains of sea animals stranded on the beach at high tide or during storms.

The three periods of the Cambrian, Ordovician, and Silurian are grouped as the first half of a longer time span called the Paleozoic era. Judging from the kinds of plant and animal life that thrived throughout the Paleozoic, the climate must have been rather mild just about everywhere. We can say this because certain fossils of the era, and even earlier, found north of the Arctic Circle differ little from those found near the Equator.

FOUR

THE AGE OF FISHES

Some 400 million years ago the Silurian gave way to the beginning of the Devonian period. During the 50 million years this period lasted, a land disturbance raised high mountains in New England, Quebec, and Nova Scotia. Another disturbance created mountains along the east coast of Australia. Large areas of North America continued to be covered by shallow seas during most of the period.

THE DEVONIAN PERIOD

The Devonian was the Age of Fishes because these animals evolved into many new forms. Some were armored, others had small scales. Some were sluggish and awkward, while others were fast, streamlined swimmers. This was a time of great change, during which two important new types of fishes evolved. One was the ancestor of sharks and rays. The other gave rise to modern bony fishes.

Another group of Devonian fishes had specialized lung sacs that enabled them to breathe air, but these fishes also had gills for breathing in the water. At the end of their fins were strong muscular lobes that could be used much like a leg. They enabled the animals to move about on land, which they most likely did when there was a need to go from a dried-up pond during times of drought to a new body of water. It was a valuable adaptation because Devonian times saw desert conditions with little rain. So some of the early land animals were fishes in search of water.

Other Devonian sea life included the builders of limestone coral reefs. Also there were many sponges, starfish, and mollusks. Some of the mollusks were cephalopods, which lived in large coiled shells or in long cone-shaped tubes. The most common cephalopod of the period was the ammonite, whose shell resembled that of the modern nautilus. In all, there have been about 10,000 cephalopod species. Today about four hundred remain, among them octopuses and squids.

The late Devonian saw the evolution of amphibians, animals such as frogs that spend part of their life cycle in water and part on land. The amphibians seem to have evolved from the lobe-finned fishes. By this time many other land organisms had appeared, including spiders and wingless insects. The slender plants of the Silurian now grew thicker and sturdier, bursting out in lush growth. Among them were horsetail rushes and tall tree ferns with stems more than 3 feet (1 meter) thick. The Devonian also brought Earth its first forests with scale trees reaching heights of 45 feet (15 meters).

A mid-Devonian seascape dating
from about 375 million years ago.

THE CARBONIFEROUS PERIOD

About 300 million years ago, during the Carboniferous period, mud deposits rich with lime settled out of many shallow seas and eventually formed today's limestone. Mountain ranges continued to be built in western Europe, and the Ouachita Mountains of Oklahoma and Arkansas were formed and re-formed. The Southern Hemisphere, meanwhile, was locked in an ice age for much of the period.

Large areas of lakes and swamps collected layer upon layer of dead trees and plants that were matted and packed down, forming peat bogs. Peat is the first stage in making coal. Over time it turned into a soft brown coal, which, with additional pressure from above for millions of years, was changed into the black coal we now burn. About half the world's usable coal was formed during the second part of the Carboniferous, mainly from giant scale trees.

Plant and animal life abounded during the Carboniferous. Ferns, club mosses, and giant scouring rushes were plentiful. There were land snails, and winged insects appeared for the first time. One species of dragonfly had a wingspan of 30 inches (75 centimeters), and there were hundreds of species of cockroaches. Splashing in the swampy pools were lungfish and king crabs, whose descendants now live only in the seas. The most interesting inhabitants of

A Carboniferous forest of some 300 million years ago.

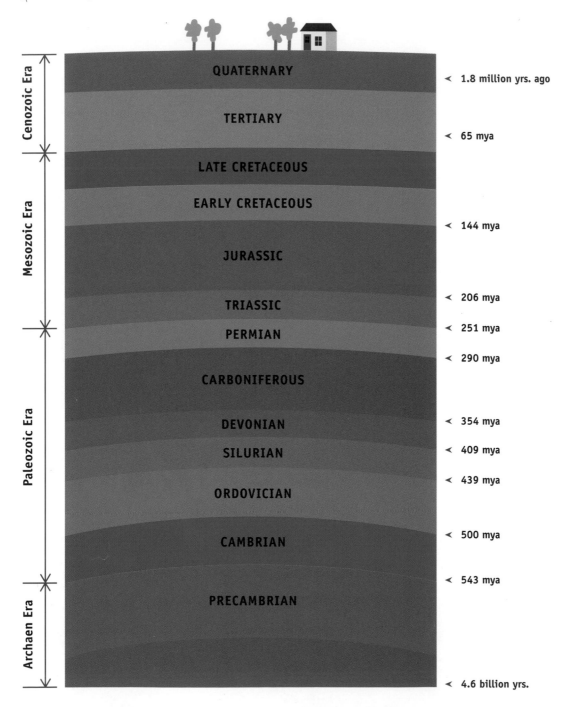

Cenozoic Era	QUATERNARY	< 1.8 million yrs. ago
	TERTIARY	< 65 mya
Mesozoic Era	LATE CRETACEOUS	
	EARLY CRETACEOUS	< 144 mya
	JURASSIC	
	TRIASSIC	< 206 mya
	PERMIAN	< 251 mya
Paleozoic Era	CARBONIFEROUS	< 290 mya
	DEVONIAN	< 354 mya
	SILURIAN	< 409 mya
	ORDOVICIAN	< 439 mya
	CAMBRIAN	< 500 mya
Archaen Era	PRECAMBRIAN	< 543 mya
		< 4.6 billion yrs.

The geologic ages of Earth span the time from Earth's formation some 4.6 billion years ago through the present. Eras are broken down into specific periods such as the Cambrian and Devonian.

these Carboniferous pools were amphibians. There were so many that this period is known as the Age of Amphibians. Some slithered around like snakes, feeding on king crabs and cockroaches. Others were the size of crocodiles and stalked the land.

Toward the end of the period one group of amphibians had evolved into reptiles. Reptiles represented an important step in evolution. Amphibians reproduce by laying their eggs in water. Frog eggs, for example, develop into tadpoles, which can survive only in water because they breathe with gills, not lungs. Only later do they develop lungs and legs, at which time they can breathe air and move about on land. But to reproduce, adult frogs must return to the water to deposit their eggs. This is true of all amphibians.

When reptiles evolved, they had an important advantage over amphibians. A reptile's eggs contain their own water supply within a watertight shell. So a baby reptile does not need a pond or puddle in order to develop. This meant that reptiles could occupy dry areas of the land where amphibians could not live and would not be in competition with them for living space and food. This tipped the scales in favor of reptiles during the next period, when life away from the water took many new turns.

THE PERMIAN PERIOD

The Permian, which began about 290 million years ago, ushered in even more change. The Appalachian Mountains south of New England were thrust up, and the Ural Mountains of Russia were formed. Along the west coast of North America volcanoes spilled lava over the land. Although the western United States was still covered by shallow seas during this period, in other parts of the Northern Hemisphere inland seas were drying up. As they did, they became more salty, making life there more difficult or impossible.

The Permian saw the development of the first seed plants—cone-bearing trees, or conifers. These were evergreen trees with slender needle-shaped leaves like today's pines and firs. Gingko trees with their distinctive fan-shaped leaves also appeared during this period. A species of gingko survives to this day.

Evidence for a general drying of the Permian land is the huge success of reptiles and a gradual dying out of amphibians. Many different kinds of reptiles emerged. Some were small, swift, and lizardlike while others were great lumbering beasts. Some ate plants, others foraged for insects, and still others consumed their fellow reptiles. Like today's reptiles, most of them were cold-blooded, meaning that their body temperature was not self-regulating, as in warm-blooded animals, but went up and down in response to the temperature of the air. However, some early reptiles were thought to be warm-blooded.

The late Permian was a time of mass extinctions among animals living in the shallow seas. Over only a few million years, at least half had disappeared. Gone were the trilobites, the ancient corals, most brachiopods, crinoids, bryozoans, and the placoderms, which had been the first true fishes.

The supercontinent of Pangaea was formed during the Permian. What today are Texas, Florida, and England were straddling the Equator in Permian times. In the Southern Hemisphere, an unconsolidated mixture of sand, pebbles, and clay sediments called till reveals how the region was

A late Carboniferous period amphibian—these were the precursors of the dinosaurs.

gripped in an ice age for about 50 million years, having begun in the Carboniferous and lasting perhaps midway into the Permian. These unassorted sediments were picked up by glaciers as they crept along. Over time, the sediments became packed and formed a rock type called *tillite*, which was released when the glaciers melted. The end of the late Paleozoic ice ages made way for a great reversal. Those parts of South America once gripped in ice were now in tropical and semitropical regions. Geologists tell us that this is important evidence that the continents have moved about in the past, just as they are drifting today.

The Permian saw the close of the Paleozoic era, which spanned some 300 million years and produced a dazzling variety of plants and animals. The two giant leaps in the progress of animal life were the evolution of amphibians from lobe-finned fish and the evolution of reptiles from amphibians. The era's passing marked the end of the Age of Ancient Life and ushered in the Age of Middle Life, known as the Mesozoic era.

FIVE

THE AGE OF REPTILES

Those terrible lizards known as the dinosaurs ruled the land for some 150 million years during the Mesozoic era. It began with the Triassic period about 250 million years ago.

The new reptiles that exploded into the Triassic in great variety and numbers were well adapted, some to water and others to desert conditions. This meant that they could spread far and wide and inhabit many parts of the environment.

THE TRIASSIC PERIOD

Seafloor spreading caused the ocean basins to change greatly during this time. Seaways continued to cover the western edges of North and South America and parts of Europe and Asia. Extensive upwelling of molten rock occurred in the New York-New Jersey area and in South America, southern Africa, Australia, and Antarctica. The impressive Dolomite Alps of northern Italy were also born. They contain abundant limestone deposits built up over millions of years. As countless tiny sea organisms died, their skeletons drifted to the seafloor, forming layer upon layer of thick limestone carpets.

Early in the Triassic, cone-bearing trees replaced the large fern and scale trees of earlier times. Cycad trees, which look like today's palms, developed during the Triassic. A climate change from warm and moist to cool and dry conditions was responsible for this change in vegetation. Invertebrates were also a common feature on the land and in lakes and streams. Among them were clams, snails, and tiny sponges. By this time,

An artist's rendering of a thecodont, a lizardlike reptile about the size of a large dog that was an ancestor of the mighty dinosaurs that roamed Earth in the Jurassic and Cretaceous periods.

The Triassic period's ancestral dinosaurs were of two main types—one with hips like those of a lizard, the other with hips like those of a bird. The bird-hip type gave rise to the duckbill dinosaurs, while the lizard-hip variety gave rise to Stegosaurus and Tyrannosaurus, both with especially powerful legs.

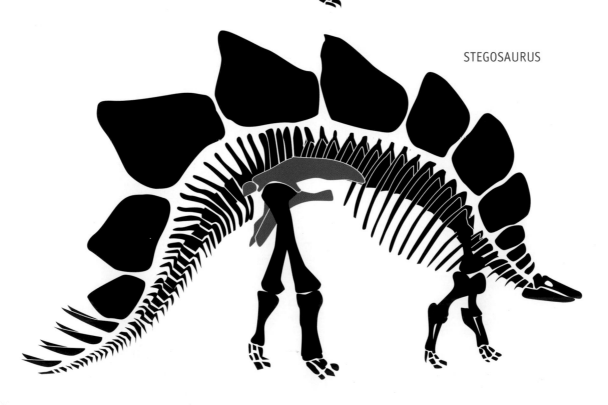

CERATOSAURUS

STEGOSAURUS

ammonites were so abundant that their fossil remains have been used to map the borders of Mesozoic seas. Crustaceans, including shrimps, were also abundant in the Triassic. Starfish, sea urchins, and sea cucumbers thrived as well.

The late Triassic's newcomers, lobsters and crabs, with their powerful arms and claws, were superior to the weaker-limbed trilobites and king crabs of earlier times. In the course of evolution, those animals that are better at competing for food generally fare better than those less equipped. That is one reason why a species may become extinct. Insects were a very successful group because they were so well adapted to the conditions of the Triassic. Most of the modern groups of insects had evolved by the end of the period.

The first mammals appeared in the late Triassic. They had evolved from mammal-like reptiles and were small and most likely covered with hair. Like later mammals, they nursed their young with milk from special glands. But unlike modern mammals, which bear live young, the first mammals most likely laid eggs as their reptile ancestors had done.

SUCCESS OF THE REPTILES

Among the new and hardier reptiles of the Triassic were turtles in their houses of bone, crocodiles, and the ocean-living plesiosaurs and ichthyosaurs. Most reptiles that took up life in the sea had to return to the land to lay their eggs, just as turtles do today. Other groups roamed the land living on a diet of grubs, snails, and insects. Thecodonts, about the size of a collie, were an outstanding evolutionary success. Unlike their slow-moving ancestors, thecodonts were swift, agile, and had large eyes well suited to spotting prey. The giant reptiles that were to come later and thunder across the landscape evolved from the thecodonts. The middle Triassic heralded their arrival.

The first-known dinosaurs were thin, two-legged animals, fossils of which have been found in Argentina and most recently in New Mexico.

Some ate vegetation while others were meat-eaters. There were two ancestral dinosaur types of the Triassic. One, called saurischians, had hips like those of a lizard and gave rise to the giant *Allosaurus* and *Tyrannosaurus rex*, which walked on two powerful legs but had only tiny forearms. The other type, called ornithischians, had hips like those of a bird. They were to give rise to the water-loving, plant-eating dinosaurs known as the duckbills. Most had thick and powerful forelegs as well as rugged hind legs. They were to later develop into *Stegosaurus* and *Triceratops*. These evolutionary marvels were to be around for millions of years.

Shrimp are among the marine organisms that have survived massive extinctions through the centuries. They started out in the Triassic, with a climate change that allowed them to thrive.

THE JURASSIC PERIOD

It was during this period, which began about 210 million years ago, that the supercontinent Pangaea broke apart into a northern half (Laurasia) and a southern half (Gondwana). The Sierra Nevada Mountains were formed. The climate turned warm and moist once again, which favored many swamps and forests, but there was no upheaval or dramatic change. In a way, the Jurassic set the stage for the great activity that was to come during the following Cretaceous period.

At the beginning of the Jurassic, North and South America had moved apart by several hundred miles. At the time North America was still attached to Africa, but as North America pulled away about 190 million years ago, a piece of Africa stuck to it and became Florida. By the end of the period, South America and Africa had begun to split apart as well.

During most of the Mesozoic, the geographic South Pole was probably located in the southern Pacific Ocean, and the North Pole was near present-day Japan. As long as the poles were in those positions, glacial ice caps on the land could not form, so the climate was generally milder than today. The gold that caused a stampede of thousands of miners to California in the mid-1800s was formed during late Jurassic times. So were the rich salt domes and sulfur deposits of the Gulf Coast.

Protozoans, microscopic animals that are only a single cell, evolved during the Jurassic and became plentiful with twenty-five major types. They include amoebas, football-shaped paramecians that spiral through the water, foraminiferans with their sticky threadlike tentacles for catching food, and radiolarians with their protective shells of sand. Modern sharks evolved during the early Jurassic and were terrors of the seas. But the true aquatic masters were the giant air-breathing reptiles called plesiosaurs, which grew to about 30 feet (10 meters). Lizards evolved during this period as did flying reptiles, the pterosaurs. The giants of the Jurassic were all plant-eaters, including *Diplodocus* and *Brachiosaurus*, both weighing 20 tons or more and measuring 60 feet (18 meters) long.

About the size of a crow, *Archaeopteryx*, the earliest known birdlike animal, evolved late in the period. Biologists search for such fossils that show how one group of animals evolves into another. There are enough examples of these "transition types"—in the case of *Archaeopteryx*, part reptile and part bird—to show the fascinating ways in which evolution has occurred and has been the driving force of biological diversity.

By the Jurassic, mammals had been around for several million years, but

An artist's impression of a Jurassic landscape. Jurassic plants include many forms still around today, including gingkos, conifers, and ferns. Fauna included ammonites and other mollusks. The dominant land animals were dinosaurs. Small mammals also existed, while birds were just beginning to appear.

57

The *Apatosaurus*, formerly known as the *Brontosaurus*, was part of the plant-eating dinosaur groups known as the sauropods. They were probably the largest animals that ever lived on land.

Below: Fish actually predated dinosaurs. As transitional amphibians, they were distinguished at that time by lobed fins.

they were not to flex their evolutionary muscles until much later. The mammals of the Mesozoic were about the size of a mouse. They lived in the undergrowth or among the branches of trees and had to rely on their swiftness to avoid being eaten by the many meat-eating reptiles.

So far in our account of the pageant of life through geologic time, we have seen that fishes evolved into amphibians, that amphibians evolved into reptiles, and that reptiles evolved into birds and mammals. Those are three very important ideas in biology, showing that life is not static but ever changing in its response to environmental change.

The *Archaeopteryx* was the earliest known animal of the bird type. It evolved late in the Jurassic.

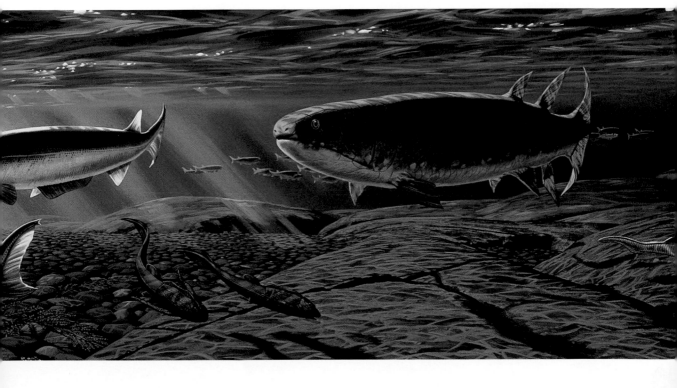

THE CRETACEOUS PERIOD

This marvelous period in Earth's history began some 145 million years ago and lasted almost 60 million years. Seas covered most of Europe, much of Asia, and nearly half of North America. The Rocky Mountains of North America and the Andes of South America were thrust up. Meanwhile, South America and Africa were splitting apart at a rate of about 1 inch (2.5 centimeters) a year. By the end of the period the two were 1,700 miles (2,800 kilometers) apart.

One of the major events of the Cretaceous was the sudden appearance some 130 million years ago of plants that produced flowers. Among them were many new types of trees—elms, oaks, maples, and others we know today. The land at this time was beginning to resemble today's Tropics. The oldest-known ants evolved late in the period. Protozoans reached fantastic numbers and developed into a wide variety of beautiful forms. The fishes we are familiar with today swam Cretaceous seas, freshwater clams made their appearance, and there were large reefs of oysters. The ammonites, so numerous earlier, were gone by the end of the period. Cretaceous plesiosaurs and giant crocodiles were joined by a newcomer, marine lizards called mosasaurs resembling mythical sea serpents.

The giant dinosaurs were such a diverse group that it would be impossible to describe them here in detail. The tyrant of the land, of course, was *Tyrannosaurus rex*, measuring up to 50 feet (15 meters) from its nose to the tip of its powerful tail, and feeding on the more passive duckbill hadrosaurs. Although many dinosaurs had tough, leathery skin, most of the land dinosaurs had bony plates of armor and spikes for protection. *Triceratops* was perhaps the most splendid of these armored monsters. It was about 30 feet (10 meters) long and plodded along on four thick legs. It had a great collar of bone with a large horn over each eye and another on the end of its nose.

Like the reptiles of today, most of the Cretaceous beasts were cold-blooded. This meant that they would be sluggish in the cool early morning hours and generally inactive until the Sun had warmed them. Then with the onset of evening, they would once again grow docile. This was not so of the newcomer mammals, who were warm-blooded and could be active at any time because of their constant body temperature. This was a great evolutionary advantage that helped tip the scales in favor of the mammals in the next geologic era.

An artist's illustration of Globegerina, a Cretaceous protozoan that survives even now.

The end of the Cretaceous also marked the end of the Mesozoic era, which had spanned some 185 million years. Like a great civilization, the dinosaurs had lived through their golden age, then fallen into decline. By the close of the Cretaceous they and the flying reptiles had all but vanished, not a trace of them to be found in sedimentary rock anywhere younger than 65 million years. The only reptiles to survive this mass extinction were the crocodiles, lizards, turtles, snakes, and the strange tuatara of New Zealand. The great science mystery that has yet to be solved is why the dinosaurs exited the scene. After a tour of the Age of Mammals, we'll explore some of the theories.

SIX

THE AGE OF MAMMALS

The Tertiary period began about 65 million years ago and ushered in the Cenozoic era, or the Age of Recent Life. It lasted for all but the past 1.8 million years and saw widespread volcanic activity in the western United States, the North Atlantic, East African, and the Mediterranean regions. During this period, the Alps and Carpathians in Europe, the Atlas Mountains in Africa, and the Himalayas in Asia were formed.

Most of the inland seas drained and evaporated from the landmasses, so by the end of the period the continents had the same general outlines that they have today. Had you been able to see them from the space shuttle during the Tertiary, you would have recognized the view as our home planet Earth.

A general cooling, coupled with the widespread extinction of many reptile species, favored the rapid evolution of mammals. By the middle of the Tertiary, most modern-day birds had evolved, as had many mammal groups including cats and whales. This period along with the next are called the Age of Mammals.

THE QUARTERNARY PERIOD

The Quaternary is the period in which we currently live. It began 1.8 million years ago and brought several major geological and biological events. Geologic activity called the Cascadian Disturbance deformed the Coast Ranges of the west coast of North America and caused widespread volcanic activity. Mounts Shasta and Rainier were formed during this time. This disturbance probably is still going on today. Massive glaciers and ice sheets spread over North America, northern Europe, and Antarctica. In places the layer of ice was 2 miles (3.2 kilometers) thick.

Several humanlike species evolved in Africa from apelike ancestors early in the period. We are the only survivors of these early humans. Modern horses also evolved around this time. The several ice ages that came and went during the Quarternary wiped out many types of mammals. But a cooling climate also favored the evolution of several gigantic mammals that were still roaming the grasslands and forests just before the last glacial ice melted about 10,000 years ago. Among them were mammoths and mastodons with their enormous curved tusks and beavers the size of a bear. Giant armadillos resembling massive armored tanks and ground sloths the

Like Mount Shasta, many of the mountains that loom large today were formed almost 2 million years ago, at the beginning of the Quaternary Period, in which we now live.

size of a rhinoceros also took their place in the order. Huge bison with horns measuring more than 6 feet (2 meters) from tip to tip grazed the grasslands and were prey for fierce saber-toothed cats. Perhaps the most dreaded predators were the short-faced bears twice the size of modern grizzlies. In all, there were more than a hundred species of these large and wondrous animals.

For some reason yet to be explained, by about 13,000 years ago most of these magnificent beasts, called *megafauna*, had disappeared. Several explanations have been offered: a marked change in climate, widespread disease, or overhunting by early native Americans. Many biologists favor the overkill theory. They believe that it would take only about a thousand years to kill off enough of the populations so that new births could no longer keep up with the

rate of deaths. Or perhaps it was a combination of factors. Whatever the cause, or causes, the disappearance of North America's megafauna marked another mass extinction.

The types of horses that we ride today evolved early in the Quaternary Period.

Gigantic mammals such as the woolly mammoth thrived during the early ice ages.

THE FIVE (OR SIX?) WORST EXTINCTIONS IN EARTH'S HISTORY

ORDOVICIAN EXTINCTION OF 439 MILLION YEARS AGO
Probably caused by a lowering of the oceans as water became locked up as glacial ice and killed off numerous marine species. Later, sea levels rose as the glaciers melted and changed the environment again. **Death toll:** 85 percent of marine organisms.

DEVONIAN EXTINCTION OF 364 MILLION YEARS AGO
Possibly caused by oxygen depletion of oceans due to asteroid impacts. **Death toll:** about 80 percent of marine life. Nothing known about extinctions of life on land.

PERMIAN EXTINCTION OF 251 MILLION YEARS AGO
Earth's worst mass extinction. A comet or asteroid strike may have been the culprit, causing massive outpourings of lava, enough to cover the entire planet to a depth of 10 feet (3 meters). But evidence for a cosmic hit has not been found. Some think the lava floods came from ruptures in the crust of what today is Siberia. **Death toll:** more than 90 percent of life.

TRIASSIC EXTINCTION OF ABOUT 215 MILLION YEARS AGO
The most likely cause was massive floods of lava pouring out of what is now the Atlantic Ocean basin. Rocks from the eruptions have been found in the eastern United States, eastern Brazil, North Africa, and Spain. Deadly global warming may have followed and contributed to the mass extinctions. **Death toll:** more than 50 percent of all species were wiped out in less than 10,000 years, a mere tick on the geological clock.

CRETACEOUS EXTINCTION OF 65 MILLION YEARS AGO

The famous extinction that is now thought to have contributed to the death of the dinosaurs. Strong evidence points to a major impact by a comet or asteroid. The impact created a gigantic crater—the Chicxulub Crater—in the floor of the Gulf of Mexico near Yucatan. Extensive lava flooding from what is now India is also suspected to have contributed to the mass extinctions. **Death toll:** up to 63 percent of marine organisms and about 20 percent of vertebrate animals living on land.

Scientists now think there may have been a total of twenty or so mass extinctions over the last billion years, all caused or helped along by comet or asteroid strikes. These catastrophic impacts may also have set the stage for a burst of new plant and animal types, as new and unoccupied ecological niches were opened and in some cases competition among surviving species was reduced. Other asteroid or comet strikes are bound to occur in the future, but exactly when no one can say.

Is there a sixth mass extinction in our future?

Most biologists agree that the world is now undergoing the fastest mass extinction of plant and animal species in its history. It is estimated that up to 27,000 species are becoming extinct every year. At that rate, half of all species could be lost in the next hundred years. The main cause for this loss of *biodiversity* is human activity. As forests are cut down for farming, lumbering, and urban development, more and more species are losing their habitats. Other human threats to species' survival are overhunting, overfishing, pollution, and climate change. In the end, this sixth extinction, which is the first to be caused by humans, may end up threatening the existence of all life on Earth, including our own.

UNTIL THE SUN DIES

Throughout Earth's geologic history, the ceaseless twisting, churning, flooding, and drying of the land; the volcanic spewing of ash, dust, and gases into the atmosphere; and the repeated grinding of massive glaciers have shaped and reshaped the face of the planet and continually altered its climate. Those geological forces have also directed the countless avenues of success and dead-end alleys that the world's stunning variety of organisms has taken. To this day evolution continues on its many paths and will do so until the Sun dies a few billion years from now. No one can say what species may still be around, or what new ones will have come and gone before the Sun burns out. But one thing is certain: Nothing will survive the fiery breaths of the dying Sun. Earth's oceans will boil away and the planet's bare crustal rocks will become too hot to support life of any kind. But that time may not be much farther into the future than the origins of life are back in the past. In the meantime, who can tell what strange and marvelous life forms will arise and enjoy their place in the Sun as we are now enjoying ours?

The surface features of Earth have been changed countless times by crustal upheavals and the eruptions of volcanic mountains such as Kilauea in Kaua'i, Hawaii.

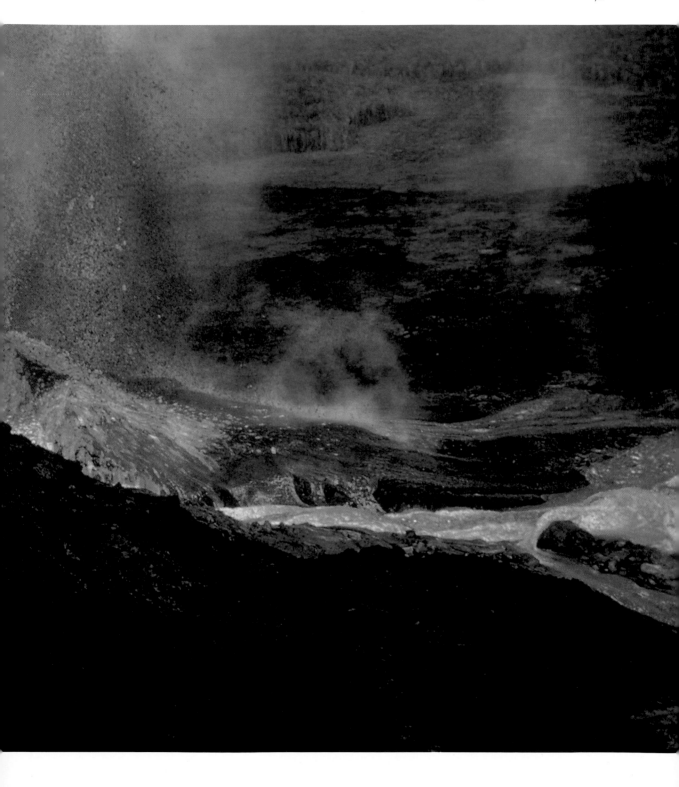

GLOSSARY

accretion During planet formation, the process in which a protoplanet sweeps up debris, in the form of planetesimals, and so accumulates mass.

amino acids Complex molecules that were among the first molecules of life some 4 billion years ago. Containing carbon, oxygen, nitrogen, and hydrogen, these molecules are the building blocks of proteins. There are about twenty different kinds of amino acids.

arthropod The animal group that includes 80 percent of all known animal species. Almost all arthropods have hard outside skeletons, jointed legs that enable them to crawl, burrow, or swim, and bodies divided into segments.

biodiversity The great variety of plant and animal life on Earth.

cell In biology, the smallest unit of living matter.

cyanobacteria Among the earliest bacteria that made their own food by combining hydrogen and carbon dioxide from the atmosphere in a process called photosynthesis. Cyanobacteria (also called blue-green algae) have survived from more than 3 billion years ago to the present.

evolution The various patterns of biological change that ultimately cause the success (adaptation) or failure (extinction) of species and produce new species of organisms.

glucose A sugar, the product of photosynthesis in green plants.

invertebrates Animals without backbones.

megafauna The diverse giant mammals, such as the woolly mammoth, inhabiting the planet just before the close of the last ice age some 10,000 years ago.

membrane A sac enclosing a cell's contents, capable of admitting certain useful molecules and expelling waste through pores.

molecule The smallest piece of an element or a compound that continues to have the same chemical and physical properties.

nebula A vast cloud of gas and dust, many of which are seen in our home galaxy and in galaxies far beyond our own.

photosynthesis The action and ability of green plants to produce glucose by combining carbon dioxide and water vapor from the air in the presence of sunlight.

planetesimals Chunks of rock, metals, and ices that were formed in the early life of the Solar System some 4.6 billion years ago and that collected in ever-larger chunks that became the planets.

protein A complex molecule manufactured in cells and that serves the cell as a nutrient, a building material, and a source of energy.

protoplanetary disk The disk of gas and dust out of which the planets and their moons condensed as the Sun was forming.

rotation Spinning around, as a planet or top rotating on its axis.

species Any one kind of animal or plant group, each member of which is like every other member in certain important ways. All populations of such a group must be capable of interbreeding and producing healthy offspring.

till An unconsolidated mixture of rocks, sand, and clay.

tillite Rock composed of till.

vertebrates Animals with backbones.

Further Reading

The following books are suitable for young readers who want to learn more about Earth's history.

Gallant, Roy A., and Christopher J. Schuberth. *Earth: The Making of a Planet*. Tarrytown NY: Marshall Cavendish Corp., 1998.

Hooper, Meredith. *The Pebble in My Pocket: A History of Our Earth*. New York: Viking Putnam Young Readers, 1996.

Smith, Norman F. *Millions & Billions of Years Ago: Dating Our Earth & Its Life*. New York: Franklin Watts, 1993.

Websites

The following Internet sites offer information about Earth's history.

http://www.extremescience.com/earthsciport.htm   This site offers a detailed description of the eras in Earth's history, as well as pictures of land and beast, and links to other sites that provide more detailed information.

http://science.nasa.gov/EarthScience.htm   This NASA site helps a visitor explore the wonders of Earth. It answers questions on how Earth is affected by various factors, including the atmosphere and pollution. Includes stories with photos and maps on various events and changes on Earth.

BIBLIOGRPAHY

Copley, Jon. "The Story of O." *Nature*. vol. 410, April 19, 2001.

Erwin, Douglas H. "The Permo-Triassic Extinction." *Nature*. vol. 367, January 20, 1994.

Gallant, Roy A., and Christopher J. Schuberth. *Earth: The Making of a Planet*. Tarrytown, NY: Marshall Cavendish Corp., 1998.

Gallant, Roy A. *From Living Cells to Dinosaurs*. New York: Franklin Watts, 1986.

Hartmann, William K., and Ron Miller. *The History of Earth: An Illustrated Chronicle of an Evolving Planet*. New York: Workman Publishing, 1991.

Jablonski, D. D. H. Erwin, and J. E. Lipps (eds.). *Evolutionary Paleobiology*. Chicago: University of Chicago Press, 1996.

Johnson, Kirk R. "Extinctions at the Antipodes." *Nature*. vol. 366, December 9, 1993.

Nisbet, E.G., and N.H. Sleep. "The Habitat and Nature of Early Life." *Nature*, vol. 409, February 2001, pp. 1083–1091.

Ryan, William, and Walter Pitman. *Noah's Flood*. New York: Simon & Schuster, 1998.

Sackmann, I. Juliana, Arnold I. Boothroyd, and Kathleen E. Kraemer. "Our Sun III: Present and Future," *The Astrophysical Journal*, 418: November 20, 1993, pp. 457–468.

Science @ NASA. "Solving Charles Darwin's 'Abominable Mystery'," April 17, 2001.

Zimmer, Carl. "How Old is…" *National Geographic*, September 2001, pp. 78–101.

INDEX

Page numbers in **boldface**
are illustrations.

ABOUT THE AUTHOR

Roy A. Gallant, called "one of the deans of American science writers for children" by *School Library Journal*, is the author of almost one hundred books on scientific subjects, including the best-selling National Geographic Society's *Atlas of Our Universe*. Among his many other books are *When the Sun Dies*; *Earth: The Making of a Planet*; *Before the Sun Dies*; *Earth's Vanishing Forests*; *The Day the Sky Split Apart*, which won the 1997 John Burroughs award for nature writing; and *Meteorite Hunter*, a collection of accounts about his expeditions to Siberia to document major meteorite impact crater events. His most recent award is a lifetime achievement award presented to him by the Maine Library Association.

From 1979 to 2000, (professor emeritus) Gallant was director of the Southworth Planetarium at the University of Southern Maine. He has taught astronomy there and at the Maine College of Art. For several years he was on the staff of New York's American Museum of Natural History and a member of the faculty of the museum's Hayden Planetarium. His specialty is documenting on film and in writing the history of major Siberian meteorite impact sites. To date, he has organized eight expeditions to Russia and is planning his ninth, which will take him into the Altai Mountains near Mongolia. He has written articles about his expeditions for *Sky & Telescope* magazine and for the journal *Meteorite*. Professor Gallant is a fellow of the Royal Astronomical Society of London and a member of the New York Academy of Sciences. He lives in Rangeley, Maine.